Real Science—Real Fun!® At Home

Pressure and Pushes and Paper Planes

For ages 8+

P9-CMQ-378

Table of Contents

Introduction

How Do Planes Fly?

Get your child interested in the science of air pressure with *Pressure and Pushes and Paper Planes*, a part of the parent-friendly *Real Science—Real Fun!® At Home* series from The Wild Goose Company! Using simple items which can usually be found around your house, you and your child can create all kinds of experiments that demonstrate the wonders of air pressure and flying. Not only will your child have fun working alone and with you to complete activities, he or she will also learn a lot about how Bernoulli's Principle gives a lift to all flying machines. Test how moving air affects surrounding objects, learn and demonstrate how air always pushes, build and fly several outstanding paper planes, and much more! And as an added bonus, many of these activities come with *More Fun Ideas to Try* which will make outstanding science fair experiments. So, roll up your sleeves and learn to fly with *Pressure and Pushes and Paper Planes!*

About This Series: Information for Parents and Other Helpful Adults

The *Real Science—Real Fun!® At Home* series makes science accessible and enjoyable for you and your child. These activity books are great for weekends, summers, science fairs, and homeschool use.

Your child can move through activities at his or her own pace. Some children will need considerable reinforcement of scientific concepts, while others will catch on quickly. But you will always be prepared to answer questions after each experiment by reviewing the *What's This All About?* section with your child. Through learning simple, fascinating explanations for each activity, your child will be amazed by the science that surrounds us.

You can easily assess how much supervision your child will need during each experiment by looking for the word *Adult* in each *Stuff You Need* list. Adults are listed as a *Stuff You Need* item any time an experiment involves using knives, matches, candles, a stove, or hot water, or cutting anything thicker than a cardboard tube. See *Lab Safety* on page 5 for more information.

Scientific Method

Be a scientist!
Most of the experiments in this book are planned for you; however, there are other ideas to try and plenty of activities you may think up on your own. Follow these steps when carrying out your own brilliant ideas for experiments.

1. Think of an idea.
When you perform an experiment from the *Real Science—Real Fun!®* At Home books, think about what you might learn from it. Or, come up with something else you want to try, and think about what you would like to learn and how you would carry out the experiment.

2. Research your own topic.
You can get some of your research from this book, but don't hesitate to ask Mom or Dad or another obliging adult to help you search the Internet or go to the library to find out more about any of the topics in this book. You may also come up with some things to research on your own. Your ideas are probably great ones!

3. Plan your experiment.
This step means deciding what materials you will need (read *Stuff You Need*), finding a good place to conduct the experiment, asking for help when you need it, and writing down the steps you will take to complete the experiment. The steps for the experiments in this book are already written for you, so if you decide to plan your own experiments, you will have examples to follow.

4. Do the experiment.
This is where you have the most fun! Roll up your sleeves and jump right in. If you did all the steps in the Scientific Method before this one, you should have a sound experiment. Just remember, if an experiment doesn't go exactly as planned, look at it as an opportunity to learn something!

5. Collect and record your data and results.
Don't forget to record what is going on. Take notes, jot down questions, and think about what is really happening and why—just like a real scientist! Some experiments in this book include an Activity Sheet where you can record information. Always check to see if you need to fill out this sheet during or after each experiment.

6. Come to a conclusion.
What did you learn from the experiment? Was the conclusion what you expected or something very surprising? Don't forget to think about how you might do the experiment differently next time.

7. Always clean up!
A clean scientist is a good scientist! Make sure your work area is clean when you leave it!

Lab Safety

Have scientific fun the right way!

Some of these activities can be dangerous if you don't follow all the directions. **Make sure you and a parent or another adult read this page before you get started!**

1. Prepare your work area.

Some of these activities are a bit messy. Others will need a sink or lots of space for you to do them. Make sure that you have the space needed to complete the activities. For example, if you need to use the sink, plan to work in a kitchen or bathroom. If you are using water, dirt, or anything else that may spill, cover your work space with newspaper. Taking steps like these will mean less cleanup time afterward!

2. Gather materials before beginning any experiment.

Nothing is more frustrating than starting an experiment and realizing that you are missing something. Read the *Stuff You Need* list for your experiment before you start, then get everything together. You may want to store your materials (except for liquids) in a box so you won't have to gather them each time.

3. Always have an adult nearby.

Having an adult around is very important. Some activities use knives, candles, or other things that can be dangerous. And sometimes, you may just need some extra help. Be sure to ask for it!

4. Wear goggles when necessary.

OK, they aren't the best fashion accessory, but safety goggles (available at school supply stores, hardware stores, and some toy stores) are required when working with anything chemical, flammable, or sharp.

5. Work safely with chemicals.

Some experiments may use household chemicals. Wear goggles when using them, and read all warning labels BEFORE you open a container. If you get any chemicals on your skin or in your eyes, flush with water. Call a physician if necessary. When you are finished, make sure chemicals are disposed of properly like it tells you on the container.

6. Don't eat or drink during experiments.

When you are concentrating on an experiment, you should never eat or drink. You could contaminate your experiment with food, or worse, poison your food with the experiment.

7. Only use fire under an adult's supervision.

If you cannot have an adult present, do not do any experiment that involves candles or matches. Know where to locate and how to use a fire extinguisher. If you catch your clothes on fire, stop, drop, and roll!

8. Have fun, but take your labs seriously.

Being rough or careless with equipment or with other people during science experiments can cause accidents. Have fun using this book, but be careful and pay attention at all times.

9. Always clean up!

Leave your work area clean! Small children or pets can swallow small objects or chemicals if you don't put everything away. So, KEEP ANYTHING DANGEROUS AWAY FROM LITTLE KIDS AND YOUR PETS! Chemicals and water can also damage carpet and furniture.

Materials List

adult
aquarium or large, clear container
balloons (4)
books (thick) (at least 4)
bowl or cup
candle (votive)
drinking glasses (large and small)
drinking straws (18)
egg (hard-boiled)
eyedropper
food coloring
funnel
hair dryer
hammer
hot pads (oven mitts)
index cards (2)
jar (small-mouth, like an olive jar)
jar (wide-mouth, like a pickle jar)
matches (wooden)
milk jug (1-gallon)
modeling clay

nail
napkin
newspaper
paper (8½" x 11")
paper towels
pie tin
push pin
saucepan
scissors
soda bottle (2-liter with cap)
soda bottle (16-ounce, glass)
soda cans (empty) (2)
soup can (empty, no label)
spool
stove top
string (18 inches long) (2 pieces)
table tennis ball
tape
tongs
tubing (rubber, 2 feet)
water

Air Pressure

There is air all around us.
You might not realize it, but air is piled on top of the Earth, and that same air pushes down and around all things. We call this **air pressure.**

To start, we need to prove that there is air pressure, and to do that, we need to understand how it got here in the first place. In the illustration to the right, we see that we are surrounded by the atmosphere, a "sea of air." (Later in this book, you will find out why "sea of air" is a really good description.) The atmosphere extends away from the Earth's surface to a distance of 50 miles or so on all sides. (Traces of the atmosphere have been found as high as 100 miles.)

This air pushes downward, creating pressure on all things.
What this really means is that there are several miles of air pushing down on Earth. As you travel higher in altitude, the air pressure drops because there is actually less air on top of you.

Air pressure changes all the time and can be measured.
This air pushes on, under, and around all things with a force of 14.7 pounds per square inch at sea level. It's a little less if you are at a high altitude. Air pressure is changed by heating the air, cooling it, changing how much of it there is, or causing it to move over a surface. You'll investigate all of those things in this book!

Don't be fooled.
Your job is to understand that there is air pressure, and that differences in air pressure can push things up, down, and over. It's important to remember that air pressure only pushes, because you will be tempted to guess that things are sucked into containers. But air is always pushed.

Smooshed Milk Jug

Do you think you can smoosh a milk jug without using your hands or feet? Do you think you could smoosh it just using air? Maybe . . .

Stuff You Need

milk jug (1-gallon)
modeling clay
tubing (rubber,
  2 feet)
water

Here's What to Do

1. Grab your milk jug and fill it to the top with water.

2. Insert the tubing into the jug. Use the clay to seal the opening around the tube, securing it in the mouth of the jug.

3. Flip the jug upside down over a sink or bathtub and watch what happens.

What's This All About?

Air is everywhere—all around us. The air that is piled on top of Earth pushes down and around all things. This is called **air pressure**. In this activity you should have seen how air pressure can push things around.

Filling the jug with water pushed out all the air that was inside. When the jug was flipped upside down, **gravity** pulled the water out of the jug. (Gravity is the force that pulls things toward Earth.) As the water came out, there was nothing to replace it. The tubing and clay kept air from getting in the jug as the water was flowing out. This made the air pressure inside the jug become lower, but the air pressure on the outside of the jug stayed the same. With the pressure on the inside of the jug being lower and the pressure on the outside being higher, the difference in pressure caused the jug to smoosh.

This activity really results in a smooshed milk jug. Hope you didn't use one you wanted to keep! If you did, have your child show you how you might "unsmoosh" the jug.

More Fun Ideas to Try

1. Do the same activity with a temperature change. Fill a 2-liter soda bottle halfway with water. Make sure the water is warm, but not hot! Cap the bottle and place it in the refrigerator. Check every five minutes to see what happens. Can you explain how the bottle changes shape? Do you know why?

2. Get adult help with this activity. Add a small amount of water to a soda can. Place the can on a hot plate and heat until the water boils. Then, use tongs to remove the can and flip it upside down into a bowl of ice water. Wow! Check out that crushing action!

Activity Sheet

Draw a picture of the milk jug while the water is still in it. Then draw a picture of the jug after the water has been emptied out of it.

Before

After

The Impossible Balloon

Air is all around us, taking up lots of space. This may be hard to imagine since you can't really see air or feel anything solid when you reach out to touch it. Try this activity and you'll see just how pushy air can be!

Stuff You Need
balloon
soda bottle
(2-liter)

Here's What to Do

1. Take your balloon and put it inside the bottle. Be careful not to drop it in the bottle. Hold on to the mouth of the balloon and pull it back over the mouth of the bottle so that it stays in place.

balloon

soda bottle

2. Put your lips on the bottle. Try to blow up the balloon.

What's This All About?

When the balloon is wrapped around the mouth of the bottle, it seals the bottle. No air can get in or out of the bottle. As you try to blow up the balloon, it pushes against the air trapped inside the bottle. The air does not like this, so it pushes back on the balloon, refusing to let it get any bigger. Air takes up space and can push back on things that push on it.

More Fun Ideas to Try

1. Try different sizes of bottles to see if you can blow up the balloon in a bigger bottle.

2. Try round balloons or long balloons to see if that makes a difference. Write down what you think might happen before you try the experiment.

3. Try a plastic bottle with a small hole punched in the bottom and see if you can blow up the balloon.

Activity Sheet

Draw an X in the circle beside the description that best describes the results of the experiment:

○ The balloon inflated to the size of a football.

○ It inflated just a little bit. It was hard to see any difference.

○ It didn't inflate at all, no matter how hard I tried.

You have learned something about the **properties** (characteristics) of air. Draw an X in the circle beside the statements about air that are true.

○ 1. Air takes up no space at all.

○ 2. Air pushes on everything.

○ 3. There is no air in the bottle.

○ 4. If there were a hole in the bottle, it would be possible to blow up the balloon.

Answers
#2 and #4 are true. #1 and #3 are false.

Kissing Cans

Here is a way to move objects with your breath! It will, uh, take your breath away!!

Here's What to Do

1. If the straws have wrappers, remove them.

2. Line them up on a table about a half inch apart (look at the drawing for help).

3. Place the empty soda cans on the straws a few inches apart.

4. Blow between the cans. If nothing happens, move them slightly closer together.

Stuff You Need
drinking straws (15)
soda cans (empty) (2)

What's This All About?

The faster air moves, the less time it has to stop and push on things. As the air between the cans increases in speed, the air pressure between the cans decreases.

The air on the outside of the cans is not moving as fast, so it has time to push a little harder than the air between the cans. As a result, the cans bump into each other.

More Fun Ideas to Try

1. See how far apart you can separate the two cans and still get a reaction.

2. Use a hair dryer and see if there is any difference.

3. Start with the two cans touching and figure out a way to get them apart.

Activity Sheet

Draw a picture of what happened to the two cans. Add arrows to show which way they moved. Make sure you include a drawing of yourself blowing between the cans!

Try the third activity under *More Fun Ideas to Try*. Show what you had to do to get the cans to separate after you blew them together.

Cartesian Diver

Ever wonder how a submarine can surface and dive? You can make a toy that will give you an idea of how it works.

Stuff You Need
eyedropper
soda bottle
(2-liter with cap)
water

Here's What to Do

1. Fill your 2-liter bottle with water.

2. Fill the eyedropper two-thirds full with water. Then, put it in the bottle and tightly screw on the cap.

3. Squeeze the sides of the bottle and see what happens to the eyedropper. Relax the pressure on the sides of the bottle, then see what happens to the eyedropper.

What's This All About?

Watch the eyedropper closely as you squeeze the sides of the bottle. Do you see how the air inside the eyedropper gets squashed? When this happens, the eyedropper fills up with water and gets less buoyant, or more dense. When its density increases, the eyedropper sinks to the bottom of the bottle.

When you let go of the sides of the bottle, the air goes back to the place where it is most comfortable. The eyedropper becomes more buoyant, or lighter, and rises to the top of the bottle again.

More Fun Ideas to Try

1. Try to figure out a way to get the eyedropper to stay right in the middle of the bottle.

2. Try placing rubber bands around the 2-liter bottle to hold the air in. See if you can find the right number of rubber bands to put around the bottle to make the eyedropper go to the bottom.

3. Try putting different amounts of water in the eyedropper. See if you can get the eyedropper to sink by itself, without squeezing. Also, try to make it so that the eyedropper will not sink no matter how hard you squeeze.

4. Try doing the experiment without the cap on the bottle. What happens?

5. Make challenging games with two or more droppers inside a 1- or 2-liter bottle. For example, you could use rubber bands to attach a hook to one eyedropper and a loop to the other (use unfolded paper clips for hooks and loops). Try to hook the droppers together by squeezing on the bottle to move them around.

6. Make a strength tester by calibrating several droppers differently so that each one takes a slightly harder squeeze to make it sink. Number the droppers, with 1 being the easiest squeeze.

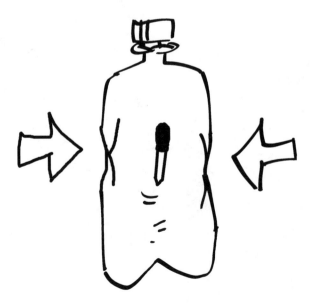

Upside-Down Water

Amaze everyone by turning a glass full of water upside down without spilling. What's the secret? A simple index card! How does it work?

Stuff You Need
drinking glass
  (any size)
index card
water

Here's What to Do

1. Do this over a sink or tub, or outside.

2. Fill the glass with water.

3. Put the index card on top of the glass and place your hand over it. Then, turn it upside down over a sink or tub.

4. Move your hand away after two seconds.

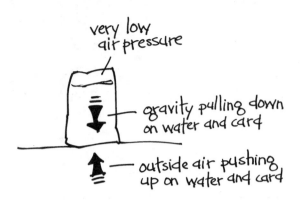

very low
air pressure

gravity pulling down
on water and card

outside air pushing
up on water and card

What's This All About?

As you flip the glass, the water inside the glass isn't pushing down on the card as hard as the air pressure that is pushing up on the card. The air pressure is exerting more force. If you wiggle the card a little bit before you move your hand, the water molecules on the card and the rim of the glass will act as a seal by sticking to each other.

More Fun Ideas to Try

1. Use different amounts of water inside the cup. How much and how little water will still give you good results?

2. Try different kinds of paper besides the index card, like construction paper, notebook paper, larger and smaller cards, and even a playing card.

3. See how long you can hold the card without the water falling out.

4. Experiment with containers of different sizes and shapes. Is a small, round glass the best container to use? What about a tiny juice glass or a great big tumbler?

5. Measure the area of the card you are using for this activity (a 3" x 5" index card would be 15 square inches). Now figure out how much force the surrounding air pressure is exerting on that card (for the 3" x 5" index card—15 x 14.7 = 220.5). Over 220 pounds of air pressure is holding that little index card against the glass of water. How much do you think the water in the glass weighs? You can figure that out by weighing the empty glass and subtracting that number from the weight of the same glass filled with water. I'll bet you it's a lot less than 220 pounds! Now, figure out how tall a glass you would need to hold enough water to push away that little index card.

Shooting Water

You've probably never created a fountain before. Don't you think it's time you did? This won't be an elaborate fountain, just a tiny one.

 Help younger children punch the holes in their soup cans so they don't smash their fingers!

Stuff You Need
adult
hammer
nail
soup can (empty, no label)
tape
water

Here's What to Do

1. Have an adult use the hammer and nail to punch three holes in a column going up the can. They should be 1 inch apart.

2. Place a piece of tape over the holes. Fill the can with water.

3. Hold the can over the sink and pull off the tape.

What's This All About?

The water in the can was pushed out by two things: the air pressure above the can and the water above each hole.

Since the bottom hole had the most water pushing on it, the water from this hole shot out the farthest. The middle hole had less pressure (force pushing on it) than the bottom hole but more than the top hole, so it didn't shoot as far. The top hole had the least pressure exerted upon it, so hardly any water shot out of it.

More Fun Ideas to Try

Experiment to see if you can make the water shoot out farther. Use additional cans, try different sizes and numbers of holes, and place them in different places on the can. Which design works best?

Activity Sheet

If you did the *More Fun Ideas to Try* activity, draw a picture of your most successful water "fountain" design. To see which design shoots water the furthest, stand over a dry, flat surface like a sidewalk or sheets of newspaper, and measure the wet patch that each design makes when you test it. After you have a winner, draw your design below, and label the drawing with your measurements.

Napkin Preserv

Do you believe it's possible to dunk a napkin in an aquarium and have it be dry when you pull it out? Doubtful, huh? See if your napkin "sinks or swims" while proving that air takes up space!

Stuff You Need

aquarium or
  large, clear
  container
drinking glass
  (any size)
napkin
water

Here's What to Do

1. Make sure the napkin is dry.

2. Stuff the napkin into the bottom of the glass. Turn the glass upside down over the floor to make sure the napkin won't fall out.

3. Keep the glass upside down and slowly lower it straight into the aquarium until the napkin and glass are both completely underwater. Don't tilt the glass at all. Now, remove the glass from the water.

4. Look at the napkin closely. Is it still dry?

What's This All About?

This activity shows that air certainly does take up space. As the glass is lowered into the container, the air inside the glass displaces (or pushes away) the water in the container. Because the water is displaced, the napkin remains dry. Pretty cool, huh?

More Fun Ideas to Try

If you are having a hard time seeing how air takes up space, just put your hands on your chest, fingers spread. Inhale, hold your breath, and then exhale. Can you see now how air takes up space?

The Fountain Bath

You might want to go outside and wear a rain hat for this one!

Stuff You Need

drinking straw
modeling clay
soda bottle
  (2-liter with cap)
water

Here's What to Do

1. Fill the bottle three-fourths full with water. Wrap a lump of clay around the middle of the straw.

2. The more air you have in the bottle, the better this experiment will work. Also, make sure that the bottom end of the straw is underwater. Insert the straw into the bottle. Use the clay to make a seal between the outside of the straw and the inside of the bottle opening.

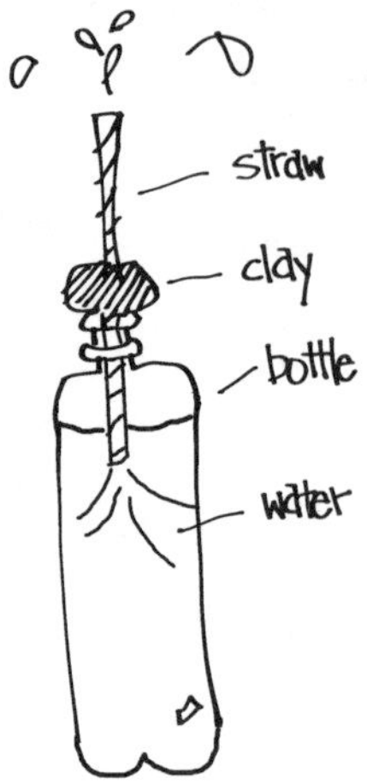

3. Blow as much air as possible into the bottle. Keep blowing until you can't force any more air into the bottle. Then remove your mouth, which releases the pressure on the bottle. Water should shoot out the top of the straw, at least 12 inches. A big breath can get the fountain close to 2 feet high as long as your nose doesn't get in the way.

(diagram labels: straw, clay, bottle, water)

What's This All About?

Air is made up of tiny things you can't see called **molecules**. Air pressure is caused by the air molecules whacking into things. If the number of air molecules increases in one place, you'll have more molecules to whack into things, and the air pressure there will also increase. If the number of air molecules decreases in one place, the air pressure will also decrease because there are fewer molecules to whack into things.

By blowing air into the bottle, you are increasing the number of air molecules in the bottle. That, of course, makes the air pressure on the inside of the bottle increase.

When you take your mouth off the bottle, the higher air pressure inside the bottle pushes the water out toward the lower air pressure outside.

More Fun Ideas to Try

1. Predict whether the size of the bottle will make a difference. Try 1-, 2-, and 3-liter bottles to test your predictions.

2. Place a horizontal straw beside the top of the vertical straw and blow air through it, across the opening of the vertical straw, to illustrate how an atomizer works. (An atomizer is like a perfume sprayer.) See the *Thirsty Straw* activity on page 42 for another example of how an atomizer works.

Vertical Flush

Can you figure out what makes water flow up into a glass in this next experiment?

Stuff You Need

adult
candle (votive)
drinking glass
 (large)
food coloring
matches
pie tin
water

This experiment uses matches, and one of the *More Fun Ideas to Try* activities suggests using other nonflammable liquids in place of water. Make sure you supervise your child during this entire experiment!

Here's What to Do

1. Put just enough water in the pie tin to cover the bottom so that there are no dry spots.

2. Put a few drops of food coloring in the water and swirl it around. The food coloring just looks cool. If you do not have it, the activity will still work.

Always use extra caution when using matches or candles in experiments.

3. Place the candle in the middle of the pie tin and light it. Place the glass over the candle and see what happens.

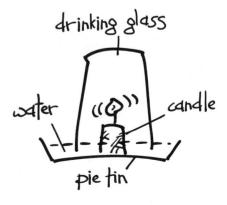

What's This All About?

You already know that one way to change air pressure is to change the number of molecules in a closed space. The presence of more molecules means higher pressure, and the presence of fewer molecules means lower pressure. A second way to change air pressure is to heat the molecules so they move faster and whack into things harder. If you cool the molecules down, they slow down. Heating air up tends to increase the air pressure; cooling it down tends to decrease the air pressure.

When you first place the glass over the burning candle, two things happen. First, the candle heats the air in the glass, which increases the air pressure by speeding up the molecules. Then, the flame takes oxygen out of the air, which changes the number of molecules and reduces the air pressure inside the glass. These two things cancel each other out, so nothing happens as long as the candle is burning.

When the candle goes out, no heat will be left to increase the air pressure in the glass. You will be left with low pressure inside because the flame has used up the oxygen molecules. The higher air pressure outside the glass then pushes the water up toward the lower air pressure inside the glass.

More Fun Ideas to Try

1. Try several different jars, larger and smaller than the first one. Make some predictions about how changing the size of the jar will affect the time it takes for the candle to burn out.

2. Try hotter and colder water to see if it makes any difference in the time it takes for the candle to go out.

3. Try one of these liquids instead of water: vinegar, dish soap, cola, milk, or juice.

DO NOT use any other liquids besides the ones suggested above. Many household cleaners and other liquids, and even some household beverages can be flammable.

Hungry Jar

Have you ever seen a jar eat anything? You are going to witness this amazing feat while doing the following activity.

Stuff You Need

adult
balloon
jar (wide-mouth,
  like a pickle jar)
matches
paper towel
water

This experiment uses matches. Make sure you supervise your child during this experiment!

Here's What to Do

1. Fill the balloon with water so that it is almost twice as big as the jar opening. Tie a knot in the end of the balloon.

2. Put the balloon on top of the jar and try as hard as you can to push the balloon into the jar with your hand flat. No squeezing! It won't go in because there is air inside the jar pushing back on the balloon.

Always have adult supervision and watch what you're doing when using matches! Make sure you drop the paper towel into the jar!

3. Take the balloon off the jar. Set it on the table. Loosely twist a paper towel into a tube. Fold the tube in half, light one end on fire, and drop it in the pickle jar. Make sure it is at the bottom of the jar.

4. While the paper towel is still burning, place the balloon on top of the jar and observe what happens.

What's This All About?

In this experiment, air pressure pushes a big water balloon inside a little pickle jar. But because of the principles of air pressure, just the right conditions are needed to succeed. As with *Vertical Flush*, two different things happen. The heat from the burning paper increases the air pressure inside the jar, but the flame also removes oxygen from the air inside the jar, reducing the air pressure inside.

You might also have seen that the water balloon jumps around at first. That's because the higher air pressure from the heat pushes it upward. Every time the balloon jumps up, though, some air leaves the jar, which lowers the pressure inside.

The grand finale (or great finish) of this activity comes when the flame goes out and you're left with fewer air molecules and a lower air pressure inside the jar. The higher air pressure outside the jar then pushes the balloon in! Wow, the jar really was hungry. It ate the balloon!

More Fun Ideas to Try

How can you make the balloon come back out of the jar using what you know about air pressure? Check the *Eggs-tra-ordinary* activity on page 28 for the solution.

Eggs-traordinary!

Did you like the last air pressure experiment? If you did, maybe you'll like this one just as much. It starts out the same but has a different twist for an ending. It's also an interesting way to eat breakfast, if your mom's not picky about you playing with your food!

Stuff You Need
adult
egg (hard-boiled)
jar (small-mouth,
    like an olive jar)
matches (wooden)

This experiment uses matches. Make sure you supervise your child during this experiment!

Here's What to Do

1. Peel a hard-boiled egg and place it on top of a clean, empty olive jar. Without breaking the egg, try your best to squish it into the jar. Bet you can't get it in there! It won't go in because the jar is full of air, which takes up space and doesn't like getting squished. So, it squishes back. If you think this sounds a lot like the last experiment, you're exactly correct!

2. Take the egg off the jar. Stick three wooden matches in one end of the egg.

Always have adult supervision and watch what you're doing when using matches! Be careful when lighting the matches and when turning the egg upside down on the jar.

3. Light the matches. Quickly place the egg on the bottle. Be sure the matches are inside the bottle or this experiment won't work. Observe the reaction. Gulp! Another hungry jar!

4. You now have a different problem. How do you get the egg OUT of the jar? Take a deep breath, tip the jar up so that the egg rests against the mouth of the jar, and blow quickly into the jar. The musty, old, smells-like-burned-paper egg will pop into your mouth. Fun, huh? And it tastes really good, too.

~~~~~~

What's This All About?

The egg got into the jar exactly the same way the balloon did in the *Hungry Jar* experiment. It has to do with molecules, air pressure, and stuff like that.

You can get the egg out of the jar because, when you blow into the jar, the burst of air increases the air pressure. The difference in air pressure pushes the egg back out of the opening. You might be able to do this experiment a couple of times, but the egg tends to fall apart from all this pushing and pulling.

Friendly Bottle

If you've never been kissed by a bottle, you don't know what you're missing. Grab the stuff in your *Stuff You Need* list and see how you like being kissed by a bottle!

Stuff You Need

adult
hot pads (oven mitts)
saucepan
soda bottle (16-ounce, glass)
stove top
tongs
water

You will need to supervise the heating process. Make sure the water doesn't boil and the bottle doesn't get too hot to touch your child's cheek.

Here's What to Do

1. Fill the pan with water and heat it until it is near boiling. Boiling is 100°C or 212°F—so don't let it get that hot! Place the soda bottle in the pan and heat it for about three minutes.

2. Use tongs to take the bottle out of the water. Use a hot pad (oven mitt) to hold the mouth of the bottle next to your cheek so that all parts of the rim are lightly touching your skin.

3. What happens next? Smooch!

© The Wild Goose Company WG 3033

Pressure and Pushes and Paper Planes

What's This All About?

When you heat the bottle, the air inside moves faster and escapes from inside the bottle. This creates an area of lower pressure inside the bottle. Pressing the bottle against your skin creates a seal. Because the inside of your mouth is at "room pressure," probably 14.7 pounds per square inch, the pressure inside your mouth is greater than the pressure inside the bottle. The difference in pressure pushes your skin a little bit inside the bottle.

More Fun Ideas to Try

Try placing an empty, uncapped 1- or 2-liter soda bottle in the freezer for a half hour. Take it out and quickly lay a quarter on the top to cover the opening. (It helps to wet the quarter first so it makes a good seal.) Watch the quarter start to dance as the air inside the bottle warms up and escapes.

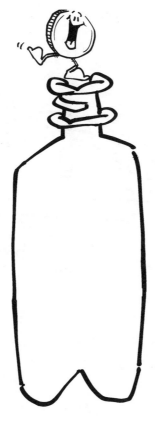

Bernoulli's Principle

Bernoulli was a Swiss mathematician and scientist who messed around with air pressure. The theory is fairly simple to understand and even easier to illustrate with the activities in this section. He discovered that the faster a fluid travels over a surface, the less it pushes on that surface. Air behaves just like a fluid, so there is a direct correlation. Check out the *Paper Tent* activity on the next page for a full explanation.

Just what this means for airplanes is illustrated below. The shape of an airplane wing is critical to the ability of the plane to generate lift and fly. The distance across the top of the wing must be longer than the distance on the underside of the wing. When the wing is traveling through the air at a fast enough speed, the air moving over the wing is moving quite a bit faster than the air moving under the wing. The faster the air travels, the less pressure it exerts on the wing. If the air traveling under the wing is moving slower, it exerts more pressure. This difference in pressure is called **lift**.

And now, the experiments . . .

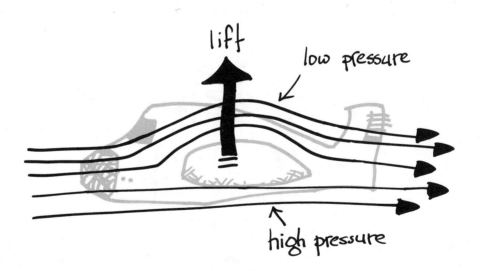

Paper Tent

The faster the air moves along any surface (such as the inside surfaces of a paper tent), the lower the air pressure it creates. This idea was discovered by Swiss mathematician Daniel Bernoulli, and it is called **Bernoulli's Principle**!

Stuff You Need
paper (8½" x 11")

Here's What to Do

1. Fold the sheet of paper in half, making a paper tent. Set up the tent using the illustration below as a guide.

2. Try to predict what will happen when you blow into the tent. Ask some friends to tell you what they think will happen. It will either stay the same, the sides will inflate and the tent will appear to get larger, or it will bend down toward the table.

3. Try this a few times to see if the results vary.

side view

What's This All About?

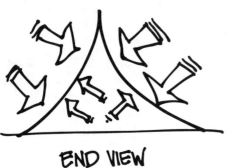

END VIEW

By blowing through the tent, you changed the air pressure inside the tent. The fact that the tent collapsed or fell down shows that you decreased the air pressure inside the tent. Did the air pressure get lower because you blew the molecules out of the way? Nice try, but nope. Your breath replaced those molecules as fast as it blew them out. Did the temperature decrease somehow? Not likely, because your breath is probably warmer than the room air. What you experienced was—ta da!—the Bernoulli Principle.

Look at it like this: imagine that you have a room jam-packed with kids, and the kids are running all over the place, bouncing into the walls. If the kids were air molecules, they would be responsible for the air pressure in the room. This is what still air does when you are not blowing inside the tent.

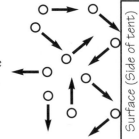

Now imagine that half the kids don't bounce around randomly, but start running toward the front of the room. Not as many kids will run into the side walls. Result: the air pressure exerted on the side walls is greatly reduced, just like the pressure on the tent walls when you blow through the tent. Make sense?

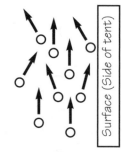

More Fun Ideas to Try

Wad up a small piece of paper and place it in the neck of a soda bottle (hold the bottle horizontally). Now, try to blow the paper into the bottle. Sounds easy enough, right? After you have tried for a while, catch your breath, then use the information you learned about air molecules to figure out why you can never, never blow that paper into the bottle!

The Flying Sheet

So, how do planes fly? Try blowing on a sheet of paper to find out.

Here's What to Do

1. Hold a sheet of paper to just under your bottom lip like the kid in the drawing. Curve the top of the paper slightly. What do you think will happen to the paper if you blow down and across the top of it? Do you think it will hit you in the chest, stay exactly where it is, or bounce up and hit you in the nose?

2. Write down your predictions, and experiment!

Stuff You Need
paper (8½" x 11")

What's This All About?

By blowing across the top of the sheet of paper, you cause many of those air molecules to move across the sheet, rather than moving all around randomly like they do in still air. So this means there's lower air pressure on top, and the higher air pressure below pushes the paper up.

More Fun Ideas to Try

The next time you take a shower, notice what the shower curtain does. Does it bulge outward from the shower area or does it try to lean in closer to you? Do you know why?

Card Bridge

How are your bridge construction skills? Let's see if you can build a bridge between two books with an index card. Then let's see how your bridge DE-struction is. Can you blow the bridge down?

Stuff You Need
books (thick) (2)
index card

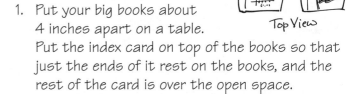

Here's What to Do

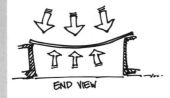

Top View

1. Put your big books about 4 inches apart on a table. Put the index card on top of the books so that just the ends of it rest on the books, and the rest of the card is over the open space.

2. What will happen when you blow between the two books and underneath the card? Either the card will remain unchanged, it will fly off the books, or it will bend down toward the table.

What's This All About?

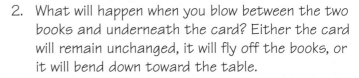

END VIEW

Blowing below the card creates a low pressure area. The card gets pushed, not sucked, downward.

More Fun Ideas to Try

What happens to the cloth top on a convertible car as it goes down the road is different. As the car moves down the road, the air inside the car is still, but the air outside the car is zipping across the top. The air inside is at a higher pressure than the air outside. The top is pushed upward by the air inside the convertible.

Sticky Papers

When wind blows through a window, do the curtains blow toward each other, away from each other, or stay where they are?

Stuff You Need
newspaper
scissors
tape

Here's What to Do

1. Cut two strips of newspaper, each about 1 inch wide and 12 inches long. Tape the strips to the edge of a table, 4 inches apart, so they hang down toward the floor.

2. What do you think will happen when you blow between the two strips of paper? They may stay the way they are, blow apart and away from each other, or come toward each other and bump or "stick" together. What do you think they'll do?

What's This All About?

FRONT VIEW

Blowing between the paper strips creates an area of low pressure. The still air outside the strips has a higher pressure and pushes them together. Is this what you thought would happen?

More Fun Ideas to Try

Try using two cheerleader pom-poms in place of the two single strips of paper. Place a fan on a table. Stand in front of the fan and hold the pom-poms (pom-side-down) about four inches apart in front of it. Make sure the fan is at the same height as the middle of the pom-poms. Which way do they blow?

Kissing Balloons

This activity is like *Sticky Papers* (page 37). It just uses balloons instead of paper! Who would've guessed it?

Stuff You Need

balloons (2)
string (18 inches long) (2 pieces)

Here's What to Do

1. Blow up the balloons. Tie the ends closed. Then, attach a string to each one. Have someone hold the two balloons at the same height or tie them to something so that they will hang free.

2. What do you think will happen when you blow between the two balloons? They will either stay the same, blow apart and away from each other, or come toward each other and bump or "kiss."

What's This All About?

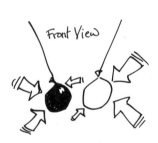

Front View

Once again, air pressure is hard at work, doing what you wouldn't expect. Most people would probably say that the balloons would fly apart, but this isn't the case. Because of the area of low pressure that you create between them, they move close together.

More Fun Ideas to Try

1. What would happen if the balloons were filled with water?

2. What would happen if the balloons were filled with helium?

Funnel Fun

This activity uses a funnel and a table tennis ball. You are to predict how the table tennis ball will react to air being blown on it. See how well you can predict the result.

Stuff You Need
funnel
table tennis ball

Here's What to Do

1. Put the funnel in your mouth like the picture. Place a table tennis ball inside the funnel and guess what will happen if you blow into the funnel. Either the ball will stay where it is, it will pop out of the funnel, or it will be pushed deeper into the funnel.

SIDE VIEW

2. Once you have done this, try flipping the funnel upside down. Find a way to keep the ball inside without using fingers, tape, glue, or any other adhesives.

What's This All About?

These two activities demonstrate that the ball will not do something you think it should. Air completely surrounds the ball, just like it surrounds every other object on Earth. As you blow into the funnel, the air right above the opening of the funnel and below the ball moves much faster than the rest of the air around the ball.

still air up here means higher air pressure

fast-moving air here means lower air pressure

This means that the ball will be pushed toward the opening where the air is moving at all times. This happens whether the funnel is upside down or right side up.

Air Ball

Another table tennis ball activity?! Since this activity is like *Funnel Fun*, what do you think will happen when air is blown on the ball?

Stuff You Need

hair dryer
table tennis ball

Here's What to Do

1. Put your hair dryer on the "no heat" setting. (You don't need heat for this trick, and the hair dryer uses less energy when it's set on cool.)

2. Place a table tennis ball into the stream of air that the hair dryer is blowing out. Let go and see what happens.

What's This All About?

The ball is held up because the air from the hair dryer pushes up and works against the gravity pulling down. How come the ball stays in the air stream instead of falling off to the side? Think about Bernoulli's Principle. The fast-moving air from the hair dryer makes the air pressure inside the stream lower than the air pressure outside, where the air is still. When the ball gets to the edge of the air stream, the higher pressure outside pushes the ball back into the lower pressure inside.

More Fun Ideas to Try

1. While the ball is floating in the stream of air, see if you can tilt it at an angle. Slowly tilt the hair dryer from side to side.

2. Try the same thing with an inflated beach ball and a leaf blower (outside, of course!).

Grabbing Spool

What kind of name for an activity is this, anyway? Can a spool hold on to a card? It doesn't have hands, but yes, it can!

Stuff You Need

bowl or cup
paper
push pin
scissors
spool

Here's What to Do

1. Use a small bowl or cup to trace a circle on paper. The circle should be about 5 inches across. Cut it out.

blow here

spool

push pin

paper circle

2. Insert a push pin in the middle of the paper circle. Put the pin and the circle directly under one hole of the spool and hold it in place (the pin should be inside the hole). What do you think will happen when you blow into the top hole of the spool and let go of the pin and paper? Will the paper circle stay in the same place, float down to the floor, or be pushed up toward the spool?

What's This All About?

As you blow down into the spool, you create a fast-moving layer of air on top of the spool and the paper circle. This causes the pressure on top of the circle to be lower, but the pressure on the bottom remains the same. This difference in air pressure pushes the paper circle up toward the spool.

More Fun Ideas to Try

Try different sizes of paper circles. Is there a point at which the circle is too large or too heavy to respond to the lower air pressure?

Thirsty Straw

What kind of a name for an activity is this, anyhow? Can a straw be thirsty? Do the following activity or ask Bernoulli!

Stuff You Need

drinking glass
  (large enough to
  hold 400 ml of
  liquid)
drinking straw
water

Here's What to Do

1. Put the straw in a glass full of water. Notice that the water level in the straw is the same as the water level in the glass.

2. Blow across the top of the straw. Will the water level in the straw stay the same, get pushed down into the water, or rise higher than the water level in the glass?

What's This All About?

In this experiment, putting a straw in the water and blowing on it created low pressure. Who would have thought you could have so much fun with air pressure?

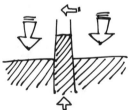

When you blow across the top of the straw, you create an area of low pressure. Bernoulli, remember? That means the atmospheric pressure (the pressure of the air) pushes the water up into the straw.

More Fun Ideas to Try

1. See *The Fountain Bath* activity on page 22.

2. Place a horizontal straw up to the top of the vertical straw and blow air through it across the opening of the vertical straw. It's the old atomizer trick again!

The Standard

This rest of this book is devoted to flying gizmos, which work thanks to air pressure! These activities will help you learn how different paper airplane designs use air pressure to fly in different ways. Fold *The Standard* design first, then move on to other designs that will challenge your folding techniques.

Stuff You Need
paper (8½" × 11")

Here's What to Do

1. Fold the paper in half vertically, then unfold it.

2. Grab one top corner and fold it down to the middle of the page. Do the same thing for the other side. You should now have something that looks like the outline of a house.

3. Fold the roof of the house completely over so that you now have a square.

4. Fold the upper right-hand corner of the house to the center line of the square. It should be about two-thirds of the way up from the bottom of the square. Do the same thing on the other side.

5. There is an odd-shaped diamond that used to be the top of the roof on the house. Fold that up so that it holds the corners of fold number 4 down.

6. Fold the airplane in half so that the folds you've been making are showing.

7. The wing is the last fold. The outside edge of the wing should be matched perfectly with the bottom of the airplane. If you want to make the plane stronger, tape the top of the wings together.

8. When you're done creating your awesome flying machine, give it a good, firm toss. If you have trouble, fold the back wings a little to help the plane fly.

1

2

3

4

5

6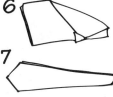

7

Jet

This is one of those planes that your mother always warned you to be careful with because it could poke someone's eye out. Find a large, open area and throw it as hard as you can.

Stuff You Need

paper (8½" x 11")

Here's What to Do

1. Fold the paper in half vertically, and then unfold it.

1

2. Grab one top corner and fold it down to the middle of the page. Do the same thing for the other side. You should now have something that looks like the outline of a house.

2

3. Grab the edges again and fold them over into the middle one more time. You now have a very steep A-frame house.

3

4. Fold the airplane in half.

4

5. The last fold is for the wings. The edge of the wing should be matched up perfectly with the bottom of the plane. If you want to make the plane even stronger, tape the wings together.

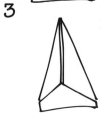

5

finished

Trickster

This plane demonstrates the movement of the airplane in response to the air that it is traveling through. If you take time to work on your design, you can really get your plane to soar like an eagle!

Stuff You Need
paper (8½" x 11")

Here's What to Do

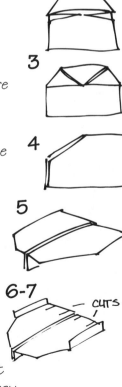

1. Fold the upper edge of the paper over to the opposite side of the paper. Unfold and repeat the same thing for the other side. You now have an X in the middle of your page.

2. This is the tricky part. Grab the center of the X and fold it into the middle so that you have made the page into another outline of a house. Use the illustrations to help you.

3. Fold the tip of the roof to the gutter.

4. Fold the airplane in half so that the folds are not showing.

5. Fold the wings down. The body of the airplane should be no more than a half inch tall.

6. Fold the outer quarter inch of the airplane wing up into the air. Tape the two wings together at the middle of the design.

7. Cut two small flaps out of the back of the wings in the sections illustrated. These will help direct the movement of the plane.

8. By bending the flaps on the back of the wing, you can get the plane to bank both left and right. If you bend both flaps the same way, you can get the plane to climb sharply into the atmosphere or crash right into the dirt.

Two Loops

There are a million ways to do this activity. Once you learn the basic design, construct five designs of your own creation. You may want to add additional loops, straws, or tail fins. You'll be amazed!

Stuff You Need

drinking straw
paper (8½" x 11")
scissors
tape

Here's What to Do

1. Cut out two strips of paper, both about 1 inch wide. One loop should be about 5 inches long, the other about 7 inches long.

2. Tape them into two loops. You should have a large loop and a small loop.

3. Place a piece of tape at each end of the unwrapped straw, so that the tape is sticky-side-up.

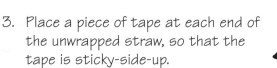

tape
straw

4. Slip the loops on either end of the straw and pat the tape down so that it holds the loops in place. The illustration should help you with this last direction.

5. Toss the plane, small loop first.

Tubular 'Copter

To fly this, all you have to do is hold it and drop it. Be sure to fly your helicopter with the rotors (flaps) pointing up. Otherwise, the helicopter may just crash instead of rotating gently to the ground. Also, if your helicopter tumbles, try snipping the ends off the rotors to make them shorter.

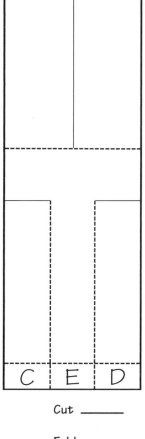

Stuff You Need
paper (8½" x 11")
scissors

Here's What to Do

1. Copy the design on the right side of the page (you can trace it or make a photocopy). Make cuts on all the solid lines, and fold on all the dotted lines.

2. Holding the design upright, fold strip A toward you and strip B away from you. Place the design on the table.

3. Fold the sides C and D of the helicopter into the middle of the design so they overlap.

4. The last fold holds the helicopter together. Take the bottom half inch or so (E) and fold it up. This is your landing gear.

A   B

C   E   D

Cut _____

Fold --------

The Wing

This is a very simple plane to make, but it's much trickier to fly. Hold the airplane between your thumb and forefinger and toss it with a gentle horizontal motion. Once you become an expert pilot, you may want to try to make different planes by making them longer or fatter or by changing the design.

Stuff You Need
paper (8½" x 11")

Here's What to Do

1. On a separate sheet of paper, make a copy of the design below and cut it out along the solid lines.

2. Accordion-fold each dotted section flat onto the next section. Tape the last fold to the paper that has not been folded.

3. Lightly crease the airplane so that it has a slight, U-shaped, upward bend.